Mostly Math

Charles Skidmore
Anne Marie Drayton

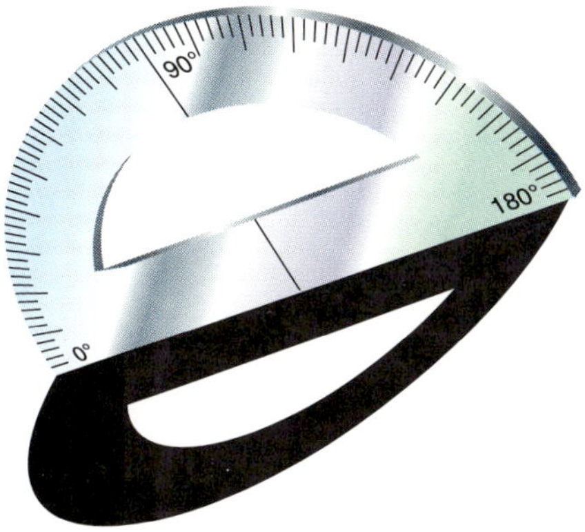

Addison-Wesley Publishing Company

CONTENTS

- 4 Photo Essay: **Tangram Tales**
- 7 Explore: Math in Social Studies
- 8 CD ROM Adventures: **Battle of the Shapes**
- 10 Explore: Math in Art
- 11 Read and Do: Create Your Own Tessellation
- 12 CD ROM Adventures: **Planet of Numbers**
- 14 Language Power: Mike's Munchies
- 16 Word Power: Math Essentials
- 18 Fable: **The Horse and the Donkey**
- 20 Play: **Two Greedy Bears**
- 24 Reading Power
- 25 Read and Do: Algebra Tiles
- 26 All Star News
- 31 Song: **The Weekend**
- 32 Amazing Facts Game

A Publication of the World Language Division

Director of Product Development
Judith M. Bittinger

Executive Editor
Elinor Chamas

Contributing Writer
Judith M. Bittinger

Editorial Development
Elinor Chamas

Text and Cover Design
Taurins Design Associates

Art Direction and Production
Taurins Design Associates

Production and Manufacturing
James W. Gibbons

Cover art: Andrew Christie
CD ROM Adventures 8–9, 12–13
　Art: Dave Sullivan
　Coloring: Paul Weiner
　Typography: Cliff Garber

Illustrators: Andrew Christie 16–17; Deborah Drummond 2; Ebet Dudley 10 *middle, bottom;* Judy Jarrett 31; Manuel King 27 *top,* 29; Sue Miller 30; Chris Reed 11; Marsha Serafin 14–15; Roni Shepherd 18–19; Jackie Snider 26; Gilles Tibo 20–23.

Photographers: © 1995 M. C. Escher/Cordon Art - Baarn, Holland. All rights reserved.10 *top;* Richard Hutchings 4–6, 28.

Copyright © 1997 by Addison-Wesley Publishing Company, Inc. All rights reserved. No part of this publication may be reproduced, stored in a retrieval system, or transmitted in any form or by any means, electronic, mechanical, photocopying, recording, or otherwise without the prior written permission of the publisher. Printed in the United States of America.

ISBN 0-201-54563-2
Mostly Math Softbound
ISBN 0-201-88542-5
Student Book 1 Hardbound (complete)
2 3 4 5 6 7 8 9 10-WC-00 99 98 97 96

Reading Corner

Try these terrific books!

Eating Fractions by Bruce McMillan
Fun with food and math.

The Doorbell Rang by Pat Hutchins
A funny story about cookies and how to divide them.

Math-a-pedia. by Randall Charles, et al.
Fun visuals in an easy-to-understand math "encyclopedia."

WARM UP

Get started with this graphing activity.

How tall are you?
How many classmates are taller? Shorter? The same?
Make a class height chart and find out.

OBSERVE AND COLLECT DATA ▼

1. Measure the height of every classmate.
2. Make a bar graph.
3. It will look something like this.

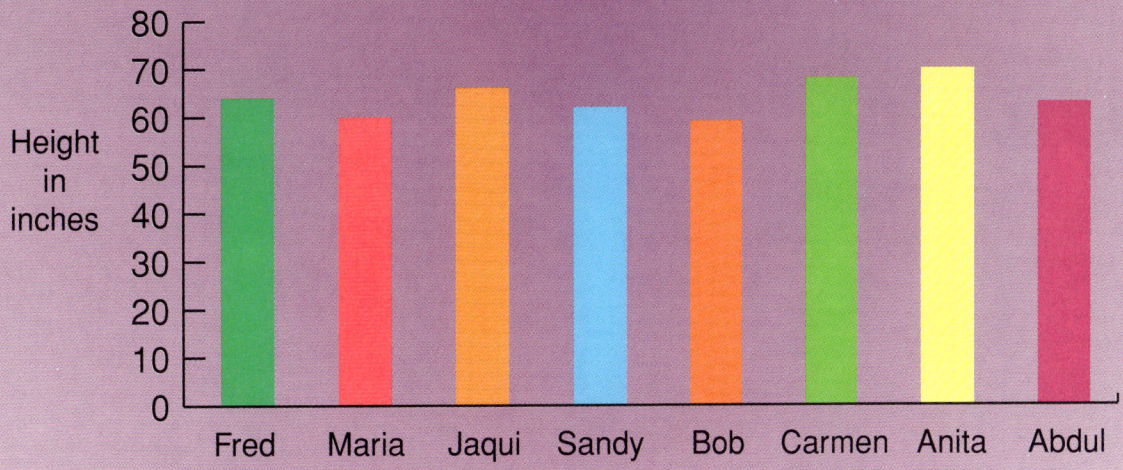

DRAW CONCLUSIONS ▼

Yes or No?
You are taller than most of your classmates.
You are shorter than most of your classmates.

All Star English

TANGRAM TALES

My brother and I like to play tangrams with our grandfather. Tangrams are a kind of puzzle. Grandfather played tangrams in China.

Grandfather tells us stories to go with the pictures. Some of the stories are about the Fox Fairy. The Fox Fairy can turn into many different animals. He can turn into a huge whale, or a little rabbit.

Fox

Rabbit

All Star English

Sometimes my brother and I make up our own tangram pictures. Then we ask my parents and my grandfather to guess what they are.

Whale

What do you think this is?

Shapes in Flags

Flags use many different shapes and colors. Most often, flags use squares, stripes, triangles, and circles. Describe these flags.

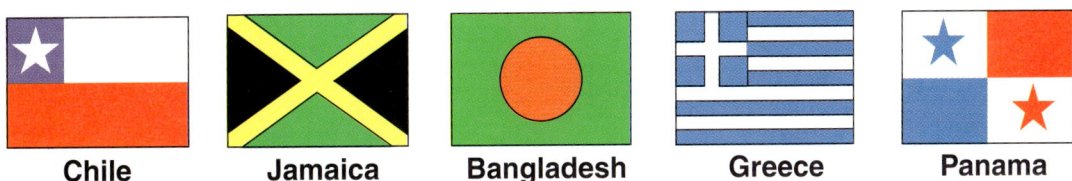

There are many other kinds of flags. The International Flag Code uses a different flag for each letter in the alphabet. Ships use the flags to send messages.

Work with a partner. Describe a flag to your partner. See if your partner can name the letter of the flag you are describing.

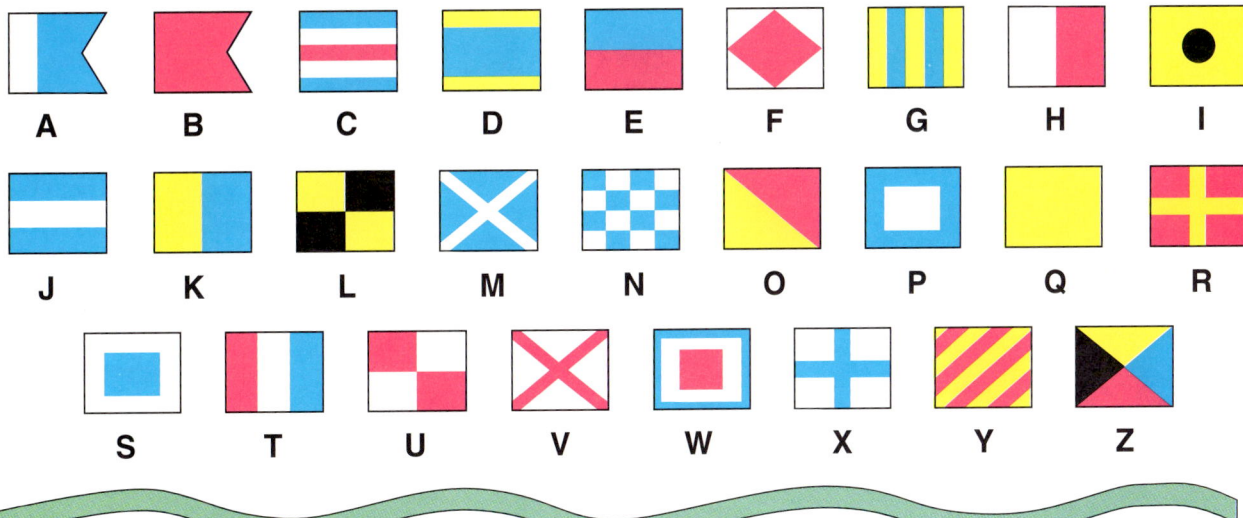

Test your flag decoding skills!
Look at each set of flags. What are the messages?

All Star English   7

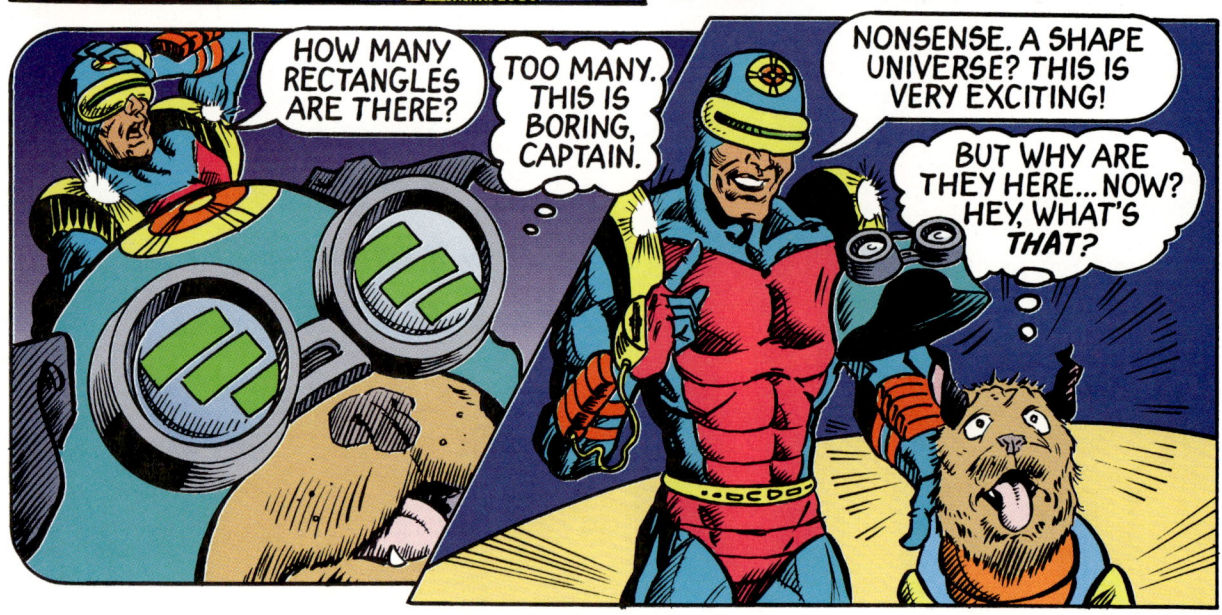

Explore... Math in Art

A tessellation is an arrangement of figures. The figures fill a flat surface. They do not overlap or leave gaps.

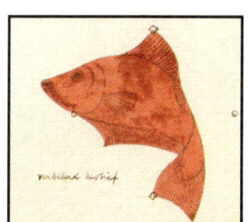

1. How many colors are there?
2. What are the colors in the drawing?
3. How many complete fish are there in the picture?
4. How many incomplete fish?
5. What shapes can you use to make a tessellation?
6. The drawing above is by a famous artist, M.C. Escher. Find out more about him.

The most common tessellation is a tiled floor.

The honeycomb of a hive of bees is also a tessellation.

10 Mostly
    ∧ Math

Art Math Music
Science Social Studies
LANGUAGE ARTS

Create Your Own Tessellation

YOU WILL NEED:
- Scissors, paper, pencil

1. Stack two sheets of paper together and fold them twice to make eight layers.

2. Draw a triangle on the top sheet of paper.

 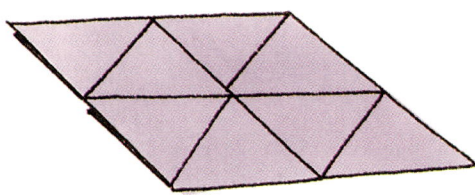

3. Cut the paper so that you get eight identical triangles.

4. Arrange the triangles so that they cover a flat surface.

Repeat the experiment with trapezoids  ,

parallelograms *, and octagons*  .

Tell which shapes do and don't make tessellations.

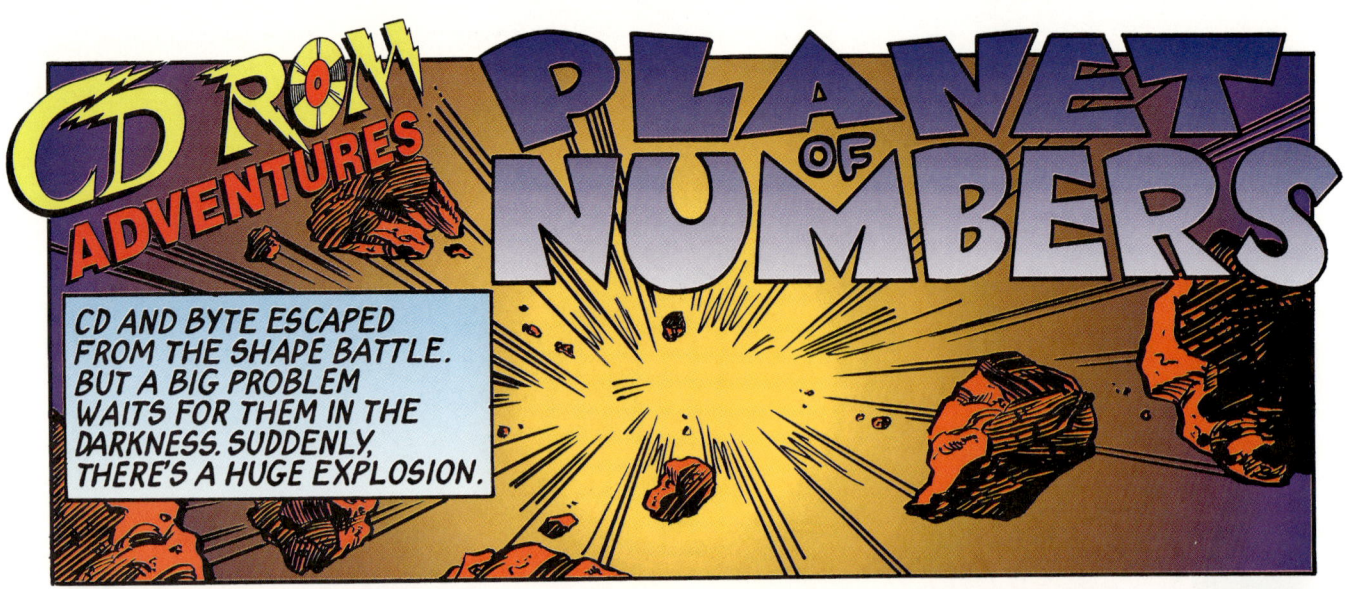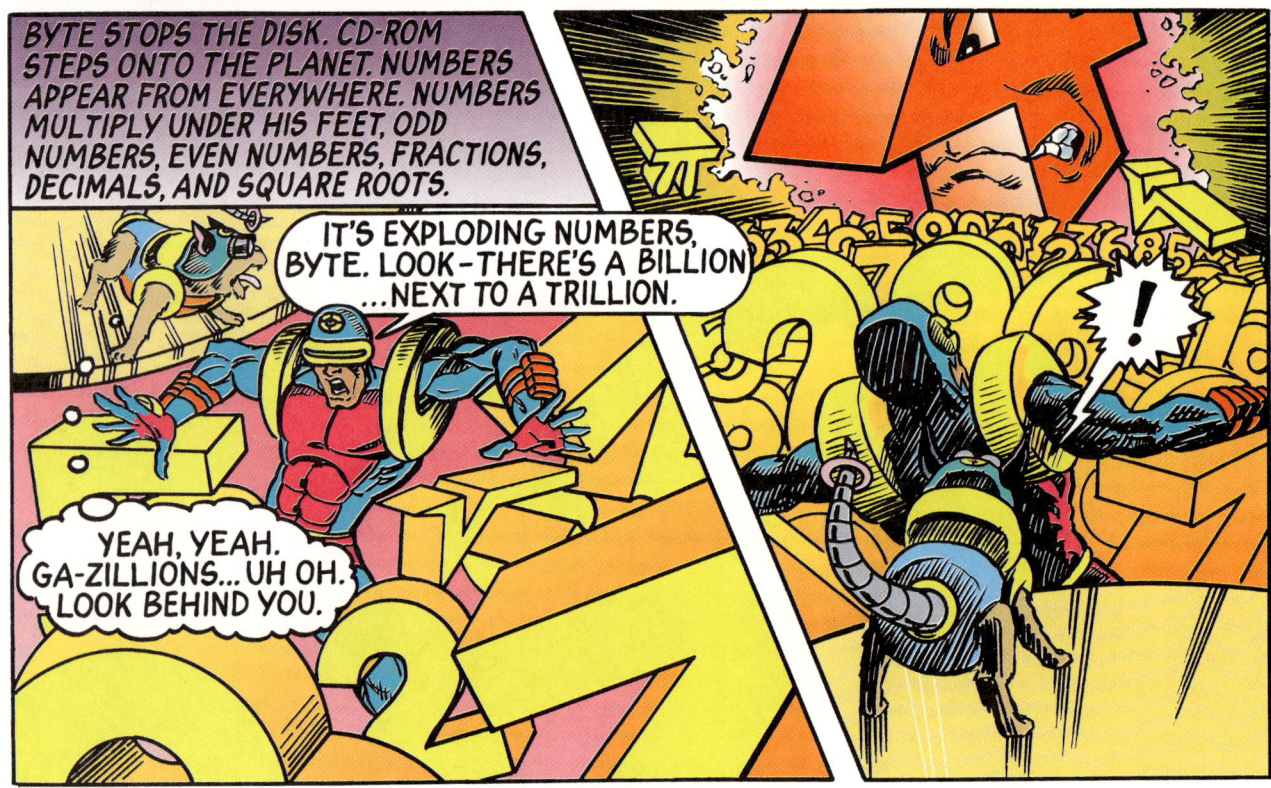

The Main Stuff

Chicken taco	1.29
Corn dog	1.49
Veggie burger	2.99
Giant salad	3.49
Buffalo burger	2.99

The Side Stuff

Cole slaw	1.19
Curly fries	.99
Plantains	1.49
Tomato salad	1.29

The Wet Stuff

	Small	Medium	Large
Soda	.79	.99	1.39
Lemonade	.79	.99	1.39
Iced tea	.79	.99	1.39
Orange Juice	1.19	1.39	1.89
Apple Juice	1.09	1.29	1.69
Milk	.59	.69	.89

The Sweet Stuff

Fruit cup	1.39
Pudding	1.39
Chocolate cake	1.39
Giant cookie	1.39

A. Answer the questions about the menu.

1. Which foods at Mike's restaurant do you like best?
2. How much does a fruit cup cost?
3. How many corn dogs can you buy with $5.00?
4. How much does a small lemonade cost?
5. How many giant cookies can you buy with $4.00?
6. How much does a veggie burger cost?
7. How much do large curly fries cost?
8. How much do all the Sweet Stuff desserts cost?
9. How much does milk cost?
10. How much does it cost to buy a giant salad, curly fries, and a large lemonade?

B. Take four friends to lunch at Mike's. Write down what each person wants. Find the cost of the meal.

C. Imagine you have $5.00. You want to buy one item from each part of the menu — the Main Stuff, The Side Stuff, The Wet Stuff, and the Sweet Stuff. What four items can you afford to buy? Write your answers on another sheet of paper.

All Star English

WORD

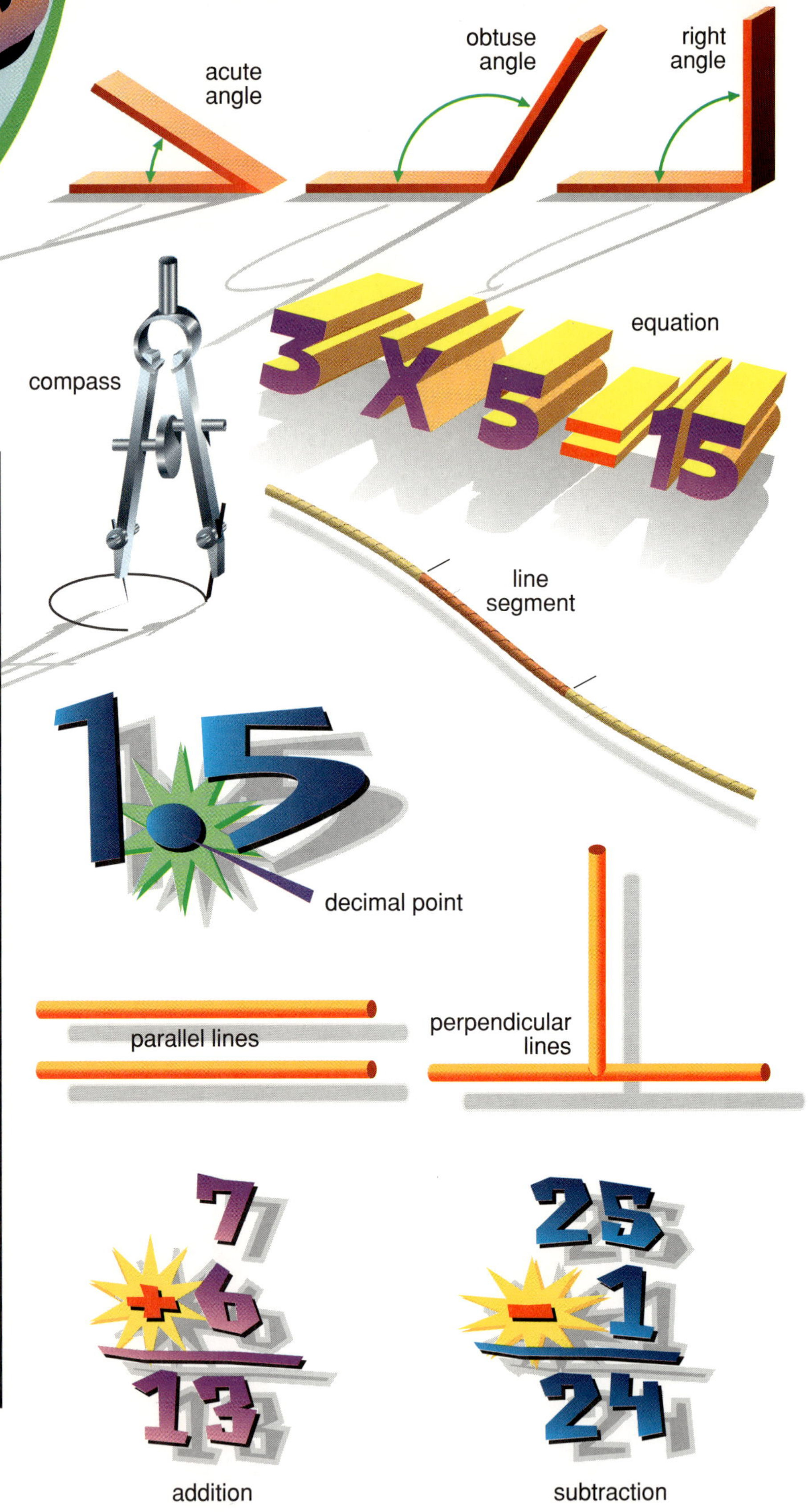

acute angle · obtuse angle · right angle

compass

equation: 3 × 5 = 15

line segment

1.5 — decimal point

parallel lines · perpendicular lines

angle space formed when two lines meet at a point

area the measure of a region, expressed in square units

calculator a machine that can add, subtract, multiply and divide

centimeter (cm) a unit of length in the metric system. 100 centimeters equal 1 meter

compass an instrument for drawing circles and measuring distance

decimal a number that shows tenths by using a decimal point

division an operation that tells how many times one number is contained in another

equation a number sentence that uses the symbol = (equal)

estimate to find an answer that is close to the exact answer

even number a whole number that has 0, 2, 4, 6, or 8 in the ones place

addition: 7 + 6 = 13

subtraction: 25 − 1 = 24

16 Mostly Math

Art · Math · Music · Science · Social Studies · LANGUAGE ARTS

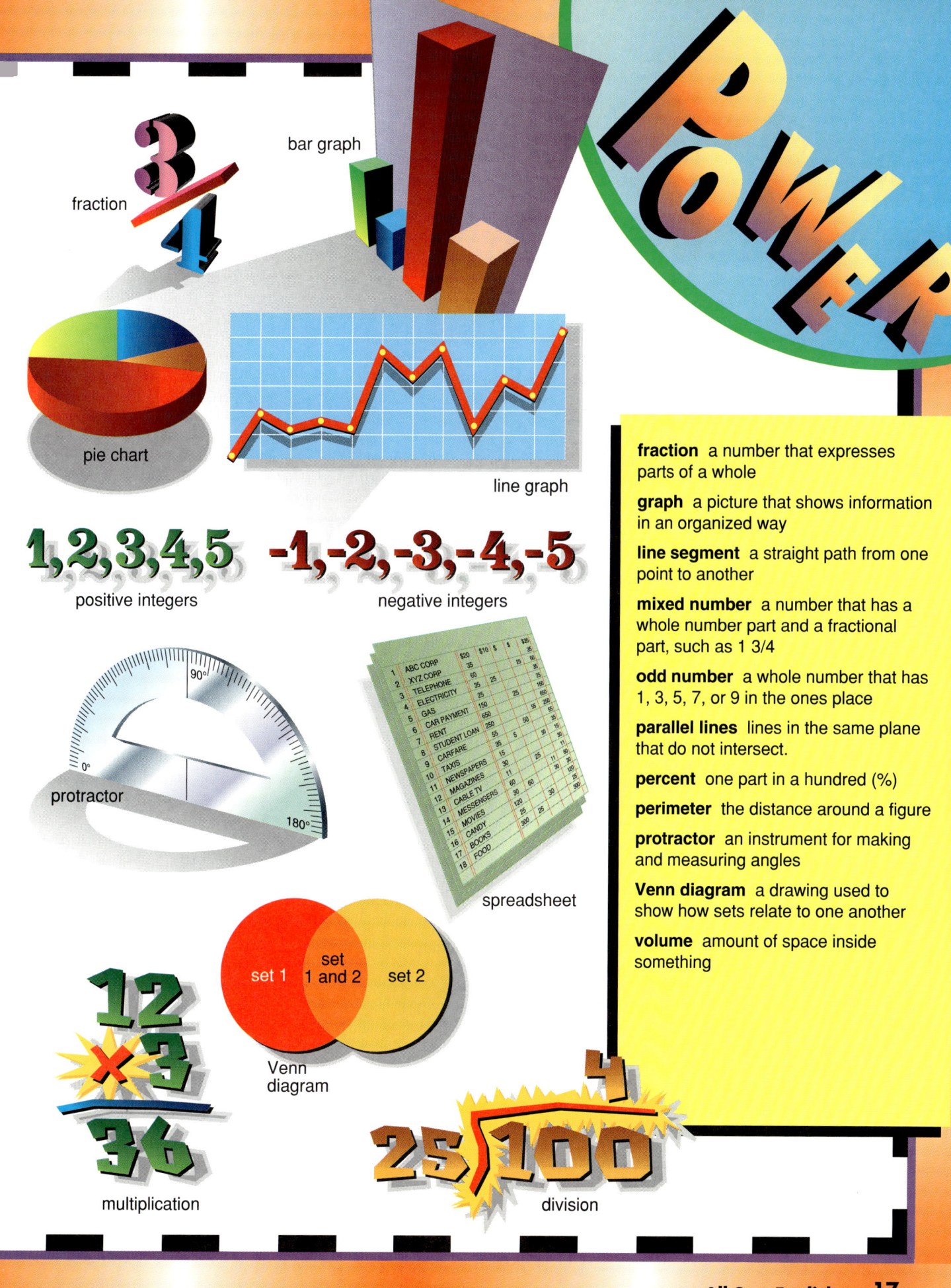

The Horse and the Donkey
A MATH FABLE

A man, a horse, and a donkey were traveling to a distant town. The man loaded all the heavy packages on the donkey's back. He put nothing on the horse's back.

After a few miles the donkey said to the horse, "I'm so tired. These packages are so heavy. Please take one half of the load."

The horse said, "No. I will not carry any packages. I am too handsome for work."

After a few more miles the donkey said to the horse, "I'm so tired. These packages are so heavy. Please take one quarter of the load."

"No," answered the horse, "I will not."

"One tenth?" asked the donkey.

"No. No. No. I am far too handsome to carry packages." Just then the donkey fell down on the ground. The man understood that the load was too much for the donkey. He put all the packages on the horse.

Moral: *The strong should help the weak.*

Two Greedy Bears
A FOLKTALE FROM EASTERN EUROPE

Characters:

                Magda   **Josef**   **Fox**

Narrator: This is a play about two young bears. One day, Magda and Josef are taking a walk. They see something by the side of the road.

Magda: What is it?

Josef: I don't know. Let's see.

Magda: Oh, it's a big, round cheese.

Josef: Yum. I'm hungry.

Narrator: The greedy bears begin to argue.

Magda: I saw it first. It's mine.

Josef : No, you saw it second. It's mine.

Magda: Give it to me.

Josef: No way - it's mine.

Magda: Look, this is silly. Let's share. You take one-half and I'll take one-half.

Narrator: Josef agrees to share. He breaks the cheese into two pieces. He looks at the two pieces, and he gives one to Magda. But, Magda isn't happy.

Magda: These pieces aren't the same size.

Josef: Yes, they are. These are two equal pieces.

Magda: No, your piece is bigger than my piece.

Josef: It's the *same*.

Narrator: Fox comes out of the woods. He is very hungry. He is also very sly.

Fox: What's the problem?

Magda: Josef's piece of cheese is bigger than mine.

Fox: Give me the cheese. I am a math professor. I can see if you are right.

Narrator: Josef and Magda give Fox the cheese. He holds one piece in each paw.

Fox: I think Josef's cheese is a little bigger. I think that Josef's cheese is 22 ounces. Magda's cheese is only 18 ounces. I can eat a little of Josef's cheese. That will make the two pieces equal.

Magda: Good idea. Eat some of his cheese.

Fox: Just a little ---just a few ounces.

Josef: That's too much, Professor. You took about 11 ounces. That was half my cheese. Now Magda has more.

Fox: I see you are a very bright young bear. Well, let me take a little of Magda's cheese.

Magda: That's too much. You ate half my cheese. Now I have only nine ounces and Josef has eleven.

Narrator: The smart fox eats some of Josef's cheese and then some of Magda's cheese. At last there is only a tiny piece for each bear.

Fox: Now your pieces are perfectly equal. They are the same—exactly one ounce each.

Magda: One ounce! That means you ate 38 ounces of cheese!

Josef: That means you ate 95% of our cheese.

Fox: Yes, that's right. And I enjoyed every percent. Next time, don't be so greedy!

A. Answer the questions about the story.

1. Where do Magda and Josef find the cheese?
2. Who sees the cheese first?
3. Why isn't Magda happy with her piece of cheese?
4. What does the Fox say he can do?
5. How does Fox plan to make the two pieces equal?
6. What percent of the cheese does Fox eat?
7. Is Fox sorry that he ate almost all the cheese?
8. Is Fox really a math professor?

B. Complete the analogies. Follow the example.

READ AND DO

Algebra Tiles

YOU WILL NEED:
- **Red paper and yellow paper**
- **Scissors**

1. Cut the paper into many small squares.
2. Think of the yellow as positive 1 (+1).
3. Think of the red as negative 1 (-1).
4. Bring together one yellow and one red square.
5. Write down the value as 0.
6. Combine groups of yellow and red squares. Tell what value results.

Work with a partner.

1. Show three different ways to have a collection of tiles that represent - 1.
2. Show three different ways to have a collection of tiles that represent 6.
3. Show three different ways to have a collection of tiles that represent - 4.

All Star English 25

Paper Money
IT'S EVERYWHERE!

Here's something hard to believe. Paper money doesn't last very long. Dollar bills wear out after about 18 months. The government gets rid of more than 7,000 tons of bills every year. They shred the bills into long strips that look like pasta. Then they shape them into bricks and dump them in landfills.

But landfills are filling up. Some companies are recycling the used-up money in very creative ways. One company takes the shredded bills and rolls them up into fireplace logs! Now people can say "I have money to burn!"

Think Tank

1. How many dollar bills would you need to fill your classroom? Here's some information to help you get started. A dollar bill is about 2.6 inches wide and 6 inches long. About 6 dollar bills cover one sheet of typing paper.

2. Two mothers and two daughters went into an ice-cream shop. They each bought one ice-cream cone. But the store clerk only charged them for three ice-cream cones. Why?

It's a Record!

The furthest a basketball has been dribbled is 265 miles in 14 days.

The record number of table tennis hits in 60 seconds is 172.

The record number of hopscotch games played by one person in one day is 307.

Hank Aaron holds the record for home runs: 755 during his 22-year career.

BULLETIN BOARD

Wanted: Used in line skates. Size: 10 or 11. Frankie, Room 68

Math Club Meeting, Wednesday 3:00 P.M., Room 10

Marching Band Practice: Monday-Wednesday-Friday: 3:30 P.M. - Playing Field

Just Joking

Where do fish keep their money?

In a river bank!

All Star English

Odd or Even Game

Children have been playing math games and counting games for thousands of years. This game comes from ancient Greece.

ODD NUMBERS
1 , 3 , 5 , 7 , 9

EVEN NUMBERS
2 , 4 , 6 , 8 , 10

Number of Players
2 or more

Object of the Game
To get as many markers as possible

Materials needed
- any small markers
  small rocks
  beans
  popcorn kernels
  (unpopped)

How to Play
1. Give each player 15 markers.
2. Have the first person hide some markers in his hand.
3. The first person turns to the second and says, "Odd or Even?"
4. The second person guesses. If the second person is right, she takes the hidden markers. If the second person is wrong, the first person keeps his markers, and play continues. Person 3 asks the next person in the circle, "Odd or even?"

AMAZING FACTS!

- George Meegan from Great Britain walked 19,019 miles from the southern tip of South America to Northern Alaska. It took him 2,426 days.

- Rick Hansen wheeled his wheelchair over 24,900 miles through four continents and 34 countries.

- The longest regularly scheduled bus route in the United States is between Miami, Florida and Los Angeles, California. The route is 2,642 miles and takes 61 hours and 45 minutes.

- The tallest man on record was Robert Wadlow. He was 8 feet 11 inches tall.

- There are more than five billion people in the world. If everybody ate a sandwich at the same time on the same day, the sandwiches would reach all the way to the moon and back!

- Astronauts can "grow" up to 2 inches when they travel in space. When they come back to Earth, they return to their normal height. What makes them shrink? Gravity!

All Star English

Poetry Corner

Addition

One plus one
is two.
Mother plus father
plus me
is three!
Sisters, brothers, plus
Grandmother Kate
is eight.

How many sisters and brothers do
I have?

Subtraction

Five minus one
is four.
My brother went
away to college,
And now
I have the bedroom
All to myself!

SONG

THE WEEKEND

Playing on the weekend, playing with my friends.
Playing on the weekend, hope it never ends.
Playing on the weekend, morning, noon or night –
Monday, it is gone.
Tuesday, it is gone.
Wednesday, it is gone
Thursday, it is gone.
Friday, it is gone.
Monday, Tuesday, Wednesday, Thursday,
Friday, they're all gone, and
It's the weekend!

EACH SQUARE =  POINTS

THREE IN A ROW =  BONUS POINTS

1 Where are tangrams from?
  a. China
  b. Korea
  c. Canada

2 Which is bigger?
  a. one half
  b. one quarter
  c. one tenth

3 Which is something to eat?
  a. menu
  b. lemonade
  c. corn dog

4 What is true?
  a. Magda was a fox.
  b. Josef was sad.
  c. The fox was smart.

5 How long does a dollar bill last?
  a. about 18 years
  b. about 18 months
  c. about 8 weeks

6 Which is the odd number?
  a. 2
  b. 3
  c. 6

7 What is not a tessellation?
  a. a tiled floor
  b. a honeycomb
  c. a piece of cheese

8 How did Rick Hansen travel?
  a. by train
  b. by wheelchair
  c. by plane

9 What makes astronauts shrink?
  a. greedy
  b. gravy
  c. gravity

Mostly Math